I0471917

U.S. Fire Administration

Matching Assistance to Firefighters Grants to the Reported Needs of the U.S. Fire Service

A Cooperative Study Authorized by U.S. Public Law 108-767, Title XXXVI

FA-304/October 2006

 Homeland Security

 NFPA

A cooperative study between:

US Fire Administration (USFA)
Directorate for Preparedness
Department of Homeland Security
and
National Fire Protection Association (NFPA)

ACKNOWLEDGEMENTS

NFPA Project Manager: John R. Hall, Jr., Ph.D.
NFPA Senior Statistician: Michael J. Karter, Jr.
USFA Project Officer: Mark A. Whitney

This second needs assessment of the U.S. fire service used the unaltered survey instrument developed in the first needs assessment, with the aim of supporting valid timelines on all questions. Once again, America's fire departments rose to the challenge, carefully reviewed their departments' capabilities and described those capabilities in forms submitted to us for use in this study.

We received essential comments at several stages from colleagues at NFPA and from the staff at USFA, and we greatly appreciate their insights.

Lastly, we want to thank the administrative personnel at NFPA, whose painstaking attention to detail and extended hours of work were instrumental in transforming a set of questions and a stack of forms into a unique database and this analysis report:

- John Baldi
- John Conlon
- Frank Deely
- Norma Candeloro
- Helen Columbo

EXECUTIVE SUMMARY

As part of the second needs assessment of the U.S. fire service, a rough comparison was made between needs reported in the first needs assessment survey and resources requested and granted to the same fire departments in 2001-2004 under the Assistance to Firefighters Grant program.

Separately, each category of need was examined to see whether needs identified in the first needs assessment survey had been reduced in the second needs assessment survey. The criterion for a reduction sufficient to justify recognition in the text was set at 5 percent or 5 percentage points (e.g., 20 percent vs. 15 percent), as appropriate. This analysis was not limited to departments that received grants but was based on all reporting departments. Note that a difference of 5 percentage points may not be statistically significant when all sources of error (e.g., sampling, non-response) are included. These comparisons are useful as indicators but should not be treated as definitive.

The following considerations should be kept in mind when examining statistics based on the 2001 and 2005 needs assessment surveys:

- These are fire department self-assessment surveys. They define needs by comparing self-reported resources to available standards or other guidance (which are identified where they are used) on what is needed to do a safe and effective job. These estimates may not be the same as would be produced by using DHS hazard/risk assessment methods or asking local authorities for their judgments of priority local needs.

- The 2005 survey was sent out shortly after Hurricane Katrina, which probably affected response rates from those areas involved.

- The response rates varied by stratum with departments protecting smaller communities responding at lower rates than those protecting larger communities. Lower response rates increase the risk for nonresponse bias in estimates. Weighting factors based on response rates and sampling fractions are used to combine results across strata.

- Results are subject to both sampling and non-sampling error. When a sample, rather than the entire population, is surveyed, there is a chance that the sample estimates may differ from the "true" population values they represent. This "sampling error" varies depending on the particular sample selected and is reflected in the "Margin of Error". In addition, the survey data are also affected by non-sampling errors, which can occur for many reasons including failure to sample a segment of the population, inability to obtain information for all respondents in the sample, the inability or unwillingness of respondents to provide correct information, and errors made in the collection or processing of the data.

The matching part of the analysis was designed to see whether the grants were effectively targeting needs. The needs trend part of the analysis was designed to see whether the grants or other actions had achieved progress in reducing needs.

The matching process is very rough and offers numerous opportunities to overstate or understate relevant needs, such as the following:

- A department could have reported a need of the general type but requested a resource of that type that is designed for a different need.

- A department could have requested and received a grant for a need of a different type than any addressed by the needs assessment report.

- A department's grant could have covered a different specific type or level of resource than what they reported having.

- Age of equipment is used to define need in some of the matching described above, but old equipment does not necessarily need replacing, and some equipment may need replacing before it reaches the age used as the threshold.

- Some needs may have arisen after the survey report was submitted or may have arisen as a result of the acquisition of other resources.

- A department may have had far more critical needs than the one(s) addressed by its grant.

For all these reasons and, no doubt, other reasons as well, this analysis can only be taken as a rough indicator of the match between needs and resources. The analysis may be useful as a basis for directing priorities in a more substantial audit or review, but it should not be used by itself as a basis for drawing conclusions.

Firefighting or EMS equipment

One-third of the matched awards and one-fourth of the granted funds for those awards went to firefighting equipment.

Overall, there was a 98 percent match for awards and a 97 percent match for award funds to some type of firefighting equipment need for this category. All six types of need showed up with high match percentages, and no one type of need dominated the others.

In the second needs assessment, the percentage of departments where there were not enough portable radios to equip everyone on a shift declined by 13 percentage points (from 77 percent to 64 percent) compared to the first survey.

None of the homeland security related needs (i.e., ability to handle any of four unusually challenging situations with local specialized equipment) showed marked improvement, nor did any of the personnel needs related to those situations.

However, there was improvement in the existence of written agreements to coordinate the use of outside personnel and equipment in a response. This is the most important step to take to improve national preparedness. The overall percentage of departments with such written agreements increased by 7 percentage points (from 19 percent to 26 percent) for the reference building collapse scenario, by 9 percentage points (from 21 percent to 30 percent) for the reference biological/chemical agent scenario, by 7 percentage points (from 33 percent to 40 percent) for the reference wildland/urban interface fire scenario, and by 5 percentage points (from 13 percent to 18 percent) for the reference flood scenario.

Lessons learned from Hurricane Katrina have a bearing on the adequacy of these agreements. "Across the emergency response community there is no common doctrine for how multiple jurisdictions should interact and respond to a single incident, or to numerous simultaneous incidents which span multiple jurisdictions. This is a critical failing." *

The continued gap in usage of a universal map reference system (the US National Grid), as documented in FA-303, *Four Years Later – A Second Needs Assessment of the U.S. Fire Service*, pp. 83-84 and Table 45, is a part of what is missing in most agreements.

---

* Hicks and Associates, Inc., *A Project Responder Report: Technology Opportunities for Implementing the National Incident Management System (NIMS)*, for the Memorial Institute for the Prevention of Terrorism and the U.S. Department of Homeland Security, October 2005.

Usage of thermal imaging cameras increased (and the need therefore decreased) by 31 percentage points (from 24 percent to 55 percent).

Personal protective equipment

More than one-third of the matched awards and granted funds for those awards went to personal protective equipment.

Overall, there was a 68 percent match for awards and a 53 percent match for award funds to some type of need for this category. The needs checked were self-contained breathing apparatus (SCBA), personal alert safety system (PASS) devices, and personal protective clothing.

Many estimated needs showed lower measures in 2005 than in 2001 in this category. The percentage of departments without enough SCBA to equip all emergency responders on a shift declined by 10 percentage points (from 70 percent to 60 percent). The percentage without enough PASS devices to equip all emergency responders on a shift declined by 14 percentage points (from 62 percent to 48 percent).

Vehicles and facility modification

Vehicles (typically engines or pumpers) represented only 8 percent of matched grants but 20 percent of grant funds for matched grants. Facility modification represented only 5 percent of matched grants and 7 percent of grant funds for matched grants.

Overall, there was an 83 percent match for vehicle awards and an 80 percent match for vehicle award funds to some type of need for this category. The sufficiency check, which showed a high need for all communities in the Needs Assessment report, accounted for most of the matching for vehicle grants.

None of the needs related to vehicles showed substantial improvement in the second needs assessment survey, and this was true for the age profile of the fleet, regardless of where the cut-off was set.

Overall, there was a 73 percent match for facility modification awards and a 70 percent match for facility modification award funds to the only type of need included in the survey for this category, namely, exhaust emission control. However, the percentage of stations not equipped for exhaust emission control did not change much, and the change varied considerably by size of community.

Training

Only 9 percent of the matched awards and only 4 percent of the granted funds for those awards went to training. This illustrates the general point that grants were sought and awarded far more for objects than for knowledge and skills. Training represented a larger share of awards and funds for larger communities than for smaller communities.

Overall, there was an 88 percent match for awards and an 80 percent match for award funds to some type of training need for this category.

Wellness and fitness programs

Only 4 percent of the matched awards and only 4 percent of the granted funds for those awards went to wellness and fitness programs.

Overall, there was a 64 percent match for awards and a 62 percent match for award funds to the need defined by not having any program of this type.

Prevention

Only 3 percent of the matched awards (Table 9) and only 2 percent of the granted funds for those awards (Table 10) went to prevention programs. This excludes the dedicated funds for national organizations and their prevention programs.

No matching with need was attempted for this category, because "prevention" covers such a broad and heterogeneous collection of programs, and the needs assessment responses are so broad and general in this area.

Prevention program usage improved in every category except arguably the two most important ones – public fire safety school education programs based on a national model and conducting fire-code inspections. The size of the population protected by departments with no plans review declined by 6 percentage points (from 29 percent to 23 percent), with no permit approval by 5 percentage points (from 45 percent to 40 percent), with no routine testing of active systems by 5 percentage points (from 49 percent to 44 percent), with no free smoke alarm distribution program by 7 percentage points (from 42 percent to 35 percent), and with no juvenile firesetter programs by 7 percentage points (from 48 percent to 41 percent).

TABLE OF CONTENTS

LIST OF TABLES AND FIGURES

Introduction

As part of the second needs assessment of the U.S. fire service, a rough comparison was made between needs reported in the first needs assessment survey and resources requested and granted to the same fire departments in 2001-2004 under the Assistance to Firefighters Grant program.

Separately, each category of need was examined to see whether needs identified in the first needs assessment survey had been reduced in the second needs assessment survey. The criterion for a reduction sufficient to justify recognition in the text was set at 5 percent or 5 percentage points (e.g., 20 percent vs. 15 percent), as appropriate. This analysis was not limited to departments that received grants but was based on all reporting departments.

The following considerations should be kept in mind when examining statistics based on the 2001 and 2005 needs assessment surveys:

- These are fire department self-assessment surveys. They define needs by comparing self-reported resources to available standards or other guidance (which are identified where they are used) on what is needed to do a safe and effective job. These estimates may not be the same as would be produced by using DHS hazard/risk assessment methods or asking local authorities for their judgments of priority local needs.

- The 2005 survey was sent out shortly after Hurricane Katrina, which probably affected response rates from those areas involved.

- The response rates varied by stratum with departments protecting smaller communities responding at lower rates than those protecting larger communities. Lower response rates increase the risk for nonresponse bias in estimates. Weighting factors based on response rates and sampling fractions are used to combine results across strata.

- Results are subject to both sampling and non-sampling error. When a sample, rather than the entire population, is surveyed, there is a chance that the sample estimates may differ from the "true" population values they represent. This "sampling error" varies depending on the particular sample selected and is reflected in the "Margin of Error". In addition, the survey data are also affected by non-sampling errors, which can occur for many reasons including failure to sample a segment of the population, inability to obtain information for all respondents in the sample, the inability or unwillingness of respondents to provide correct information, and errors made in the collection or processing of the data.

The matching part of the analysis was designed to see whether the grants were effectively targeting needs. The needs trend part of the analysis was designed to see whether the grants or other actions had achieved progress in reducing needs.

Matching process

The first step was matching the needs-assessment database with the grantee database to develop combined records on needs and grants received for a group of fire departments. The matching of fire department identifiers was done by and under the auspices of staff of the U.S. Fire Administration. NFPA staff then completed the process of creating combined records.

This combined database was then slightly reduced to only those departments that reported their population coverage in their needs assessment response. That restriction permitted analysis of the database by community size.

The result was 753 fire department matches for 2001 grants, 2,415 for 2002 grants, 3,713 for 2003 grants, and 3,276 for 2004 grants. This meant a combined total of 10,157 departments receiving grants, with some departments receiving multiple grants in multiple years.

Some grants were also multi-part (e.g., with a part for firefighting equipment and a part for training). These parts were treated as distinct grants for purposes of analysis, so that there were a total of 14,925 grants to departments in form suitable for comparison to reported needs.

The comparison required a positive match. Therefore, if a fire department submitted a needs assessment response but left all the questions related to a particular need blank, that department was counted as not having reported that type of need.

The results of the matching part of the analysis are based directly on the above database of awards and survey responses; they were not constructed to be nationally representative or representative of different size communities. It is likely that there were variations across the population strata in terms of percentage of departments that received grants, percentage of departments that responded to the needs assessment survey in 2001, and percentage of departments that qualified for the matching database because they received a grant, responded to the needs assessment survey, reported their population protected, and were successfully matched between the grant and needs assessment survey databases. No attempt has been made to weight responses in different strata to reflect these differences, when constructing statistics for all departments combined.

Limits of the grant-need matching

The matching process is very rough and offers numerous opportunities to overstate or understate relevant needs, such as the following:

- A department could have reported a need of the general type but requested a resource of that type that is designed for a different need. For example, a department might have reported a need for EMS training but have requested and received a grant for structural firefighting training while not having reported a need for that type of training.

- A department could have requested and received a grant for a need of a different type than any addressed by the needs assessment report. For example, a department might have reported a need for vehicle firefighting training, which is not one of the types of training asked about in the survey.

- A department's grant could have covered a different specific type or level of resource than what they reported having. For example, a department might have everyone trained in technical rescue – hence, no reported need at the threshold used for reporting – but not have everyone trained in technical rescue to the level required for a very challenging building-collapse situation, and that might have been the training they sought in their grant application.

- Age of equipment is used to define need in some of the matching described above, but old equipment does not necessarily need replacing, and some equipment may need replacing before it reaches the age used as the threshold.

- Some needs may have arisen after the survey report was submitted (e.g., engines reached the 30-year threshold) or may have arisen as a result of the acquisition of other resources (e.g., training is needed in the use of equipment acquired in another part of the grant or through other means).

- A department may have had far more critical needs than the one(s) addressed by its grant. For example, replacement of an old but serviceable engine might have been sought ahead of training and equipment that the department did not have at all.

For all these reasons and, no doubt, other reasons as well, this analysis can only be taken as a rough indicator of the match between needs and resources. The analysis may be useful as a basis for directing priorities in a more substantial audit, but it should not be used by itself as a basis for drawing adverse conclusions.

Parts of needs assessment survey used to identify needs by category

Here are the questions and answers used to define needs for each category of grants:

Firefighting equipment (combined with grants labeled for "EMS equipment" or "Equipment")

There were three distinguishable types of need in this category.

The first was equipment required by NFPA standards (specifically, portable radios):

Q. 27a: How many of your emergency responders on-duty on a single shift can be equipped with portable radios? Need existed if the answer was Most, Some, or None, and therefore not All. (Other needs that could have been derived from survey responses but were not included in this analysis were that (a) not all radios were water-resistant and intrinsically safe in an explosive atmosphere, and (b) there were not reserve portable radios equal to at least 10 percent of in-service radios.)

The second was equipment deemed necessary by the departments to respond to unusually challenging incidents that fell within their responsibility. These were the homeland-security related needs:

Q. 36a: Is technical rescue and EMS for a building with 50 occupants after structural collapse within your department's responsibility?

Q. 36c: If [technical rescue and EMS for a building with 50 occupants after structural collapse is within your department's responsibility], how far would you have to go to obtain enough specialized equipment to handle this incident?

Need is defined if the answer to Q. 36a is Yes and the answer to Q. 36c is Regional, State, or National and not Local Would Be Enough.

Q. 37a: Is hazmat and EMS for an incident involving chemical/biological agents and 10 injuries within your department's responsibility?

Q. 37c: If [hazmat and EMS for an incident involving chemical/biological agents and 10 injuries is within your department's responsibility], how far would you have to go to obtain enough specialized equipment to handle this incident?

Need is defined if the answer to Q. 37a is Yes and the answer to Q. 37c is Regional, State, or National and not Local Would Be Enough.

Q. 38a: Is a wildland/urban interface fire affecting 500 acres within your department's responsibility?

Q. 38c: If [wildland/urban interface fire affecting 500 acres is within your department's responsibility], how far would you have to go to obtain enough specialized equipment to handle this incident?

Need is defined if the answer to Q. 38a is Yes and the answer to Q. 38c is Regional, State, or National and not Local Would Be Enough.

Q. 39a: Is mitigation (confining, slowing, etc.) of a developing major flood within your department's responsibility?

Q. 39c: If [mitigation (confining, slowing, etc.) of a developing major flood is within your department's responsibility], how far would you have to go to obtain enough specialized equipment to handle this incident?

Need is defined if the answer to Q. 39a is Yes and the answer to Q. 39c is Regional, State, or National and not Local Would Be Enough.

The last was equipment deemed useful but not required by any standard (specifically, thermal imaging cameras):

Q. 40: Do you have any [thermal imaging cameras] now or plan to acquire any? Need is defined if any answer is given other than Now Own.

Personal protective equipment

Q. 28a: How many emergency responders on-duty on a single shift can be equipped with self-contained breathing apparatus (SCBA)? Need was defined if the answer was Most, Some, None, and therefore not All. (Other needs that could have been derived from survey responses but were not included in this analysis were that any SCBA were 10 years old or older.)

Q. 29: How many of your emergency responders on-duty on a single shift are equipped with Personal Alert Safety System (PASS) devices? Need was defined if the answer was Most, Some, None, and therefore not All.

Q. 30a: How many of your emergency responders are equipped with personal protective clothing? Need was defined if the answer was Most, Some, None, and therefore not All. (Other needs that could have been derived from survey responses but were not included in this analysis were that (a) any clothing is at least 10 years old, or (b) there were not reserve clothing to equip 10 percent of emergency responders.)

Vehicles (combined with grants for "Firefighting vehicles")

Q. 6: What share (%) of your apparatus was [each of the listed alternatives]? Need was defined if a percentage greater than zero was entered under Converted Vehicles Not Designed as FD Apparatus.

Q24e. Number of engines/pumpers in service [that are] 30 or more years old. Need was defined if a number greater than zero was entered in this blank. (Other needs that could have been derived from survey responses but were not included in this analysis were lower age thresholds of 20 or 15 years old.)

Sufficiency. There were not enough engines to equip enough stations (one engine per station), optimally located, to provide community coverage in accordance with NFPA standards and ISO formulas. This employed the formulas used in the Needs Assessment report, with different distance criteria for smaller vs. larger communities, as described in the report. Note that this was not an assessment of whether there were enough engines for the community's existing stations but rather whether there were enough engines for the number of stations required to appropriately cover the community's entire area.

Facility modification

Q23d. Number of fire stations/Number equipped for exhaust emission control (e.g., diesel exhaust extraction). Need was defined if a number greater than zero was entered in this blank.

Wellness and fitness

Q18. Does your department have a program to maintain basic firefighter fitness and health (e.g., as required in NFPA 1500)? Need was defined by a No answer.

Training (combined with grants for "EMS training")

Q13a. Structural firefighting. Is this a role your department performs?

Q13b. If yes, how many of your personnel who perform this duty have received formal training (not just on-the-job)?

Need is defined if the answer to Q13a is Yes and the answer to Q13b is Most, Some, or None, and therefore not All. (Other needs that could have been derived from survey responses but were not included in this analysis are some levels of personnel certification.)

Q14a. Emergency medical service (EMS). Is this a role your department performs?

Q14b. If yes, how many of your personnel who perform this duty have received formal training (not just on-the-job)?

Need is defined if the answer to Q14a is Yes and the answer to Q14b is Most, Some, or None, and therefore not All. (Other needs that could have been derived from survey responses but were not included in this analysis are some levels of personnel certification.)

Q15a. Hazardous materials (Hazmat). Is this a role your department performs?

Q15b. If yes, how many of your personnel who perform this duty have received formal training (not just on-the-job)?

Need is defined if the answer to Q15a is Yes and the answer to Q15b is Most, Some, or None, and therefore not All. (Other needs that could have been derived from survey responses but were not included in this analysis are some levels of personnel certification.)

Q16a. Wildland firefighting. Is this a role your department performs?

Q16b. If yes, how many of your personnel who perform this duty have received formal training (not just on-the-job)?

Need is defined if the answer to Q16a is Yes and the answer to Q16b is Most, Some, or None, and therefore not All.

Q17a. Technical rescue. Is this a role your department performs?

Q17b. If yes, how many of your personnel who perform this duty have received formal training (not just on-the-job)?

Need is defined if the answer to Q17a is Yes and the answer to Q17b is Most, Some, or None, and therefore not All.

Prevention

No matching with need was attempted for this category, because "prevention" covers such a broad and heterogeneous collection of programs, and the needs assessment responses are so broad and general in this area.

Analysis Results by Category of Need

Firefighting equipment

One-third of the matched awards (Table 1) and one-fourth of the granted funds for those awards (Table 2) went to firefighting equipment.

Overall, there was a 98 percent match for awards and a 97 percent match for award funds to some type of firefighting equipment need for this category. All six types of need showed up with high match percentages, and no one type of need dominated the others.

In the second needs assessment, the percentage of departments where there were not enough portable radios to equip everyone on a shift declined by 13 percentage points (from 77 percent to 64 percent) compared to the first survey.

None of the homeland security related needs (i.e., ability to handle any of four unusually challenging situations with local specialized equipment) showed marked improvement, nor did any of the personnel needs related to those situations.

However, there was improvement in the existence of written agreements to coordinate the use of outside personnel and equipment in a response. This is the most important step to take to improve national preparedness. The overall percentage of departments with such written agreements increased by 7 percentage points (from 19 percent to 26 percent) for the reference building collapse scenario, by 9 percentage points (from 21 percent to 30 percent) for the reference biological/chemical agent scenario, by 7 percentage points (from 33 percent to 40 percent) for the reference wildland/urban interface fire scenario, and by 5 percentage points (from 13 percent to 18 percent) for the reference flood scenario.

Lessons learned from Hurricane Katrina have a bearing on the adequacy of these agreements. "Across the emergency response community there is no common doctrine for how multiple jurisdictions should interact and respond to a single incident, or to numerous simultaneous incidents which span multiple jurisdictions. This is a critical failing." *

The continued gap in usage of a universal map reference system (the US National Grid), as documented in FA-303, *Four Years Later – A Second Needs Assessment of the U.S. Fire Service*, pp. 83-84 and Table 45, is a part of what is missing in most agreements.

Usage of thermal imaging cameras, which is not required by any NFPA standard, saw one of the largest increases in any part of the second needs assessment survey. The usage

---

* Hicks and Associates, Inc., *A Project Responder Report: Technology Opportunities for Implementing the National Incident Management System (NIMS)*, for the Memorial Institute for the Prevention of Terrorism and the U.S. Department of Homeland Security, October 2005.

increased (and the need therefore decreased) by 31 percentage points (from 24 percent to 55 percent).

This increase in usage even outpaced expressions of intent to acquire cameras, as reported in the first needs assessment survey, strongly suggesting that the availability of grant funds made the difference in these purchases.

Personal protective equipment

More than one-third of the matched awards and granted funds for those awards (Tables 3-4) went to personal protective equipment.

Overall, there was a 68 percent match for awards and a 53 percent match for award funds to some type of need for this category. The needs checked were self-contained breathing apparatus (SCBA), personal alert safety system (PASS) devices, and personal protective clothing.

Matching varied substantially by size of community as did this category's share of grants. For communities of 50,000 or more population, less than 30 percent of awards were for this category and 10 percent or less of awards showing matching with a reported need. Conversely, for rural communities, 42 percent of awards were for personal protective equipment and there was a 93 percent matching rate with need.

The gap in matching could represent grants to replace old equipment with equipment that performs better and more in compliance with NFPA standards, grants to achieve a reserve in compliance with NFPA standards, or grants addressed to types of equipment other than the three types included in the survey.

Many estimated needs showed lower measures in 2005 than in 2001 in this category. The percentage of departments without enough SCBA to equip all emergency responders on a shift declined by 10 percentage points (from 70 percent to 60 percent). The percentage without enough PASS devices to equip all emergency responders on a shift declined by 14 percentage points (from 62 percent to 48 percent). Also, the percentage of departments with some SCBA at least 10 years old declined by 11 percentage points (from 45 percent to 34 percent) and the percentage with some personal protective clothing at least 10 years old declined by 5 percentage points (from 37 percent to 32 percent).

Vehicles and facility modification

Vehicles (typically engines or pumpers) represented only 8 percent of matched grants but 20 percent of grant funds for matched grants (Tables 5-6). Facility modification is included on the same tables because the only need addressed in the survey was vehicle-related (i.e., exhaust emission control). Facility modification represented only 5 percent of matched grants and 7 percent of grant funds for matched grants.

Overall, there was an 83 percent match for vehicle awards and an 80 percent match for vehicle award funds to some type of need for this category. The sufficiency check, which showed a high need for all communities in the Needs Assessment report, accounted for most of the matching for vehicle grants.

Converted vehicles were an issue primarily for smaller communities where presumably they were used by volunteer firefighters. Vehicle grants also represented a larger share of total grants for smaller communities (e.g., 13 percent for rural communities of less than 2,500 population compared to 4 percent or less for communities of 10,000 or more population). This was even more evident as a share of total grant funds (e.g., 40 percent for rural communities compared to 10 percent or less for communities of 25,000 or more population).

The gap in vehicle matching could in part represent grants to replace old vehicles that were not 30 years old.

None of the needs related to vehicles showed substantial improvement in the second needs assessment survey, and this was true for the age profile of the fleet, regardless of where the cut-off was set.

Overall, there was a 73 percent match for facility modification awards and a 70 percent match for facility modification award funds to the only type of need included in the survey for this category, namely, exhaust emission control. There are known to be other well-established facility design needs related to firefighter safety and health (which were the only modifications these grants were intended to address), such as safety of passage between floors (replacing the old slide poles). Therefore, the 73 percent is actually a pretty high match percentage for only one type of need.

The percentage of stations not equipped for exhaust emission control declined by 6 percentage points (from 78 percent to 72 percent) in the second needs assessment survey.

Training

Only 9 percent of the matched awards (Table 7) and only 4 percent of the granted funds for those awards (Table 8) went to training. This illustrates the general point that grants were sought and awarded far more for objects than for knowledge and skills. Training represented a larger share of awards and funds for larger communities than for smaller communities.

Overall, there was an 88 percent match for awards and an 80 percent match for award funds to some type of training need for this category. EMS training showed a lower matching rate to needs (32 percent) than did any of the other four types of training needs examined (40 percent for structural firefighting training, 52 percent for hazmat response training, 55 percent each for wildland firefighting training and technical rescue training). Matching rates were low in larger communities for every type of training other than technical rescue training. Matching rates for technical rescue training were higher in larger communities than in smaller communities, while the reverse was true for the other four types of training.

So many specific elements of training are included in each of these types that one must be cautious in attributing too much meaning to the matching gaps. There is a major difference between providing all involved personnel *any* formal training and providing them will all needed training.

Wellness and fitness programs

Only 4 percent of the matched awards (Table 9) and only 4 percent of the granted funds for those awards (Table 10) went to wellness and fitness programs. Wellness and fitness programs represented a larger share of awards and funds for larger communities than for smaller communities (e.g., more than 10 percent of awards for communities of 100,000 population or more vs. 2 percent of awards for communities of less than 5,000 population).

Overall, there was a 64 percent match for awards and a 62 percent match for award funds to the need defined by not having any program of this type. There is a major difference between providing *any* program and providing a complete program with all necessary elements.

Prevention

Only 3 percent of the matched awards (Table 9) and only 2 percent of the granted funds for those awards (Table 10) went to prevention programs. This excludes the dedicated funds for national organizations and their prevention programs.

Prevention program usage improved in every category except arguably the two most important ones – public fire safety school education programs based on a national model and conducting fire-code inspections. The size of the population protected by departments with no plans review declined by 6 percentage points (from 29 percent to 23 percent), with no permit approval by 5 percentage points (from 45 percent to 40 percent), with no routine testing of active systems by 5 percentage points (from 49 percent to 44 percent), with no free smoke alarm distribution program by 7 percentage points (from 42 percent to 35 percent), and with no juvenile firesetter programs by 7 percentage points (from 48 percent to 41 percent).

Table 1
Reported Needs vs. Awarded Grants – Firefighting Equipment

Community Size	Percent of Awards	Need Any Question	Need Q27a	Need Q36a,c	Need Q37a,c	Need Q38a,c	Need Q39a,c	Need Q40
500,000 or more	32%	94%	56%	38%	16%	44%	44%	9%
250,000 to 499,999	28%	100%	50%	47%	33%	43%	30%	23%
100,000 to 249,999	39%	97%	54%	66%	40%	43%	57%	19%
50,000 to 99,999	33%	96%	50%	64%	63%	38%	45%	25%
25,000 to 49,999	33%	94%	50%	67%	60%	35%	45%	34%
10,000 to 24,999	33%	96%	54%	63%	63%	42%	39%	45%
5,000 to 9,999	35%	98%	73%	55%	56%	53%	42%	63%
2,500 to 4,999	37%	99%	78%	52%	53%	47%	37%	80%
Under 2,500	34%	100%	85%	39%	42%	49%	32%	90%
Total	34%	98%	70%	53%	53%	46%	38%	64%

Note: Reported needs defined by indicated responses to questions. Need requires positive indication of need; blank answer to question is interpreted as no need. Percents in second column are percent of all grants to departments in that population stratum for which the grant was for the indicated resource, which here is firefighting equipment. Percents in third through ninth columns are (number of grants to departments in that population stratum for the indicated resource where the grantee department reported a need on any of the specified questions)/(number of grants to departments in that population stratum for the indicated resource). Firefighting equipment accounted for 5,119 of the 14,925 grants in the database.

Q. 27a: How many of your emergency responders on-duty on a single shift can be equipped with portable radios? Need existed if the answer was Most, Some, or None, and therefore not All.

Q. 36a: Is technical rescue and EMS for a building with 50 occupants after structural collapse within your department's responsibility?
Q. 36c: If [technical rescue and EMS for a building with 50 occupants after structural collapse is within your department's responsibility], how far would you have to go to obtain enough specialized equipment to handle this incident? Need is defined if the answer to Q. 36a is Yes and the answer to Q. 36c is Regional, State, or National and not Local Would Be Enough.

Q. 37a: Is hazmat and EMS for an incident involving chemical/biological agents and 10 injuries within your department's responsibility?
Q. 37c: If [hazmat and EMS for an incident involving chemical/biological agents and 10 injuries is within your department's responsibility], how far would you have to go to obtain enough specialized equipment to handle this incident? Need is defined if the answer to Q. 37a is Yes and the answer to Q. 37c is Regional, State, or National and not Local Would Be Enough.

Table 1

Reported Needs vs. Awarded Grants – Firefighting Equipment (Continued)

Q. 38a: Is a wildland/urban interface fire affecting 500 acres within your department's responsibility?

Q. 38c: If [wildland/urban interface fire affecting 500 acres is within your department's responsibility], how far would you have to go to obtain enough specialized equipment to handle this incident? Need is defined if the answer to Q. 38a is Yes and the answer to Q. 38c is Regional, State, or National and not Local Would Be Enough.

Q. 39a: Is mitigation (confining, slowing, etc.) of a developing major flood within your department's responsibility?

Q. 39c: If [mitigation (confining, slowing, etc.) of a developing major flood is within your department's responsibility], how far would you have to go to obtain enough specialized equipment to handle this incident? Need is defined if the answer to Q. 39a is Yes and the answer to Q. 39c is Regional, State, or National and not Local Would Be Enough.

Q. 40: Do you have any [thermal imaging cameras] now or plan to acquire any? Need is defined if any answer is given other than Now Own.

Source: USFA files on Fire Act grantees for "Firefighting Equipment," "EMS Equipment," and "Equipment", and matching to USFA/NFPA Needs Assessment 2001 survey responses

14

Table 2

Reported Needs vs. Grant Amounts – Firefighting Equipment

Community Size	Percent of Dollars Granted	Need Any Question	Need Q27a	Need Q36a,c	Need Q37a,c	Need Q38a,c	Need Q39a,c	Need Q40
500,000 or more	28%	98%	60%	37%	17%	42%	48%	17%
250,000 to 499,999	20%	100%	38%	68%	34%	32%	40%	7%
100,000 to 249,999	42%	97%	49%	59%	37%	43%	54%	20%
50,000 to 99,999	33%	94%	50%	69%	61%	48%	51%	21%
25,000 to 49,999	28%	93%	51%	66%	59%	36%	44%	30%
10,000 to 24,999	25%	97%	52%	65%	65%	45%	44%	44%
5,000 to 9,999	22%	99%	72%	59%	59%	55%	44%	62%
2,500 to 4,999	21%	99%	76%	51%	55%	50%	37%	79%
Under 2,500	18%	100%	84%	39%	43%	49%	33%	89%
Total	25%	97%	61%	58%	53%	46%	43%	48%

Note: Reported needs defined by indicated responses to questions. Need requires positive indication of need; blank answer to question is interpreted as no need. Percents in second column are percent of all grant dollars to departments in that population stratum for which the grant was for the indicated resource, which here is firefighting equipment. Percents in third through ninth columns are (number of grant dollars to departments in that population stratum for the indicated resource where the grantee department reported a need on any of the specified questions)/(number of grant dollars to departments in that population stratum for the indicated resource). Firefighting equipment accounted for $237 million of the $960 million in grant money in the database.

Q. 27a: How many of your emergency responders on-duty on a single shift can be equipped with portable radios? Need existed if the answer was Most, Some, or None, and therefore not All.

Q. 36a: Is technical rescue and EMS for a building with 50 occupants after structural collapse within your department's responsibility?
Q. 36c: If [technical rescue and EMS for a building with 50 occupants after structural collapse is within your department's responsibility], how far would you have to go to obtain enough specialized equipment to handle this incident? Need is defined if the answer to Q. 36a is Yes and the answer to Q. 36c is Regional, State, or National and not Local Would Be Enough.

Q. 37a: Is hazmat and EMS for an incident involving chemical/biological agents and 10 injuries within your department's responsibility?
Q. 37c: If [hazmat and EMS for an incident involving chemical/biological agents and 10 injuries is within your department's responsibility], how far would you have to go to obtain enough specialized equipment to handle this incident? Need is defined if the answer to Q. 37a is Yes and the answer to Q. 37c is Regional, State, or National and not Local Would Be Enough.

15

Table 2

Reported Needs vs. Grant Amounts – Firefighting Equipment (Continued)

Q. 38a: Is a wildland/urban interface fire affecting 500 acres within your department's responsibility?

Q. 38c: If [wildland/urban interface fire affecting 500 acres is within your department's responsibility], how far would you have to go to obtain enough specialized equipment to handle this incident? Need is defined if the answer to Q. 38a is Yes and the answer to Q. 38c is Regional, State, or National and not Local Would Be Enough.

Q. 39a: Is mitigation (confining, slowing, etc.) of a developing major flood within your department's responsibility?

Q. 39c: If [mitigation (confining, slowing, etc.) of a developing major flood is within your department's responsibility], how far would you have to go to obtain enough specialized equipment to handle this incident? Need is defined if the answer to Q. 39a is Yes and the answer to Q. 39c is Regional, State, or National and not Local Would Be Enough.

Q. 40: Do you have any [thermal imaging cameras] now or plan to acquire any? Need is defined if any answer is given other than Now Own.

Source: USFA files on Fire Act grantees for "Firefighting Equipment," "EMS Equipment," and "Equipment", and matching to USFA/NFPA Needs Assessment 2001 survey responses

16

Table 3
Reported Needs vs. Awarded Grants – Personal Protective Equipment

Community Size	Percent of Awards	Need Any Question	Need Q28a	Need Q29	Need Q30a
500,000 or more	28%	4%	4%	4%	0%
250,000 to 499,999	27%	10%	0%	3%	7%
100,000 to 249,999	20%	7%	6%	4%	3%
50,000 to 99,999	29%	7%	4%	6%	1%
25,000 to 49,999	29%	22%	16%	15%	3%
10,000 to 24,999	34%	40%	32%	29%	4%
5,000 to 9,999	39%	70%	62%	53%	11%
2,500 to 4,999	41%	85%	79%	69%	15%
Under 2,500	42%	93%	87%	78%	23%
Total	37%	68%	61%	54%	13%

Note: Reported needs defined by indicated responses to questions. Need requires positive indication of need; blank answer to question is interpreted as no need. Percents in second column are percent of all grants to departments in that population stratum for which the grant was for the indicated resource, which here is personal protective equipment. Percents in third through sixth columns are (number of grants to departments in that population stratum for the indicated resource where the grantee department reported a need on any of the specified questions)/(number of grants to departments in that population stratum for the indicated resource). Personal protective equipment accounted for 5,533 of the 14,925 grants in the database.

Q. 28a: How many emergency responders on-duty on a single shift can be equipped with self-contained breathing apparatus (SCBA)? Need was defined if the answer was Most, Some, None, and therefore not All.

Q. 29: How many of your emergency responders on-duty on a single shift are equipped with Personal Alert Safety System (PASS) devices? Need was defined if the answer was Most, Some, None, and therefore not All.

Q. 30a: How many of your emergency responders are equipped with personal protective clothing? Need was defined if the answer was Most, Some, None, and therefore not All.

Source: USFA files on Fire Act grant recipients and matching to USFA/NFPA Needs Assessment 2001 survey responses

17

Table 4

Reported Needs vs. Grant Amounts – Personal Protective Equipment

Community Size	Percent of Dollars Granted	Need Any Question	Need Q28a	Need Q29	Need Q30a
500,000 or more	32%	4%	4%	4%	0%
250,000 to 499,999	37%	9%	0%	7%	2%
100,000 to 249,999	28%	3%	3%	1%	1%
50,000 to 99,999	32%	11%	6%	10%	0%
25,000 to 49,999	36%	21%	16%	17%	3%
10,000 to 24,999	42%	39%	30%	28%	4%
5,000 to 9,999	44%	68%	59%	52%	10%
2,500 to 4,999	43%	83%	78%	66%	15%
Under 2,500	39%	92%	86%	77%	21%
Total	39%	53%	47%	42%	9%

Note: Reported needs defined by indicated responses to questions. Need requires positive indication of need; blank answer to question is interpreted as no need. Percents in second column are percent of all grant dollars to departments in that population stratum for which the grant was for the indicated resource, which here is personal protective equipment. Percents in third through sixth columns are (number of grant dollars to departments in that population stratum for the indicated resource where the grantee department reported a need on any of the specified questions)/(number of grant dollars to departments in that population stratum for the indicated resource). Personal protective equipment accounted for $373 million of the $960 million in grant money in the database.

Q. 28a: How many emergency responders on-duty on a single shift can be equipped with self-contained breathing apparatus (SCBA)? Need was defined if the answer was Most, Some, None, and therefore not All.

Q. 29: How many of your emergency responders on-duty on a single shift are equipped with Personal Alert Safety System (PASS) devices? Need was defined if the answer was Most, Some, None, and therefore not All.

Q. 30a: How many of your emergency responders are equipped with personal protective clothing? Need was defined if the answer was Most, Some, None, and therefore not All.

Source: USFA files on Fire Act grant recipients and matching to USFA/NFPA Needs Assessment 2001 survey responses

18

Table 5
Reported Needs vs. Awarded Grants – Vehicles and Facility Modification

Community Size	Vehicle Percent of Awards	Need Any Question	Need Q6e	Need Q24e	Need Sufficiency	Facility Percent of Awards	Need Q23d
500,000 or more	1%	100%	0%	0%	100%	9%	89%
250,000 to 499,999	0%	NA	NA	NA	NA	9%	60%
100,000 to 249,999	3%	40%	0%	30%	40%	6%	48%
50,000 to 99,999	4%	71%	4%	13%	63%	7%	64%
25,000 to 49,999	3%	48%	2%	13%	44%	9%	70%
10,000 to 24,999	4%	76%	24%	30%	58%	8%	74%
5,000 to 9,999	7%	82%	33%	35%	62%	6%	77%
2,500 to 4,999	9%	85%	35%	38%	66%	3%	74%
Under 2,500	13%	89%	43%	47%	68%	2%	77%
Total	8%	83%	35%	39%	64%	5%	73%

NA – Not applicable because there were no such awards.

Note: Reported needs defined by indicated responses to questions. Need requires positive indication of need; blank answer to question is interpreted as no need. Percents in second and seventh column are percents of all grants to departments in that population stratum for which the grant was for the indicated resource, which here is vehicles and facility modification, respectively. Percents in third through eighth columns are (number of grants to departments in that population stratum for the indicated resource where the grantee department reported a need on any of the specified questions)/(number of grants to departments in that population stratum for the indicated resource). Vehicles and facility modification accounted for 1,147 and 767, respectively, of the 14,925 grants in the database.

Q. 6: What share (%) of your apparatus was [each of the listed alternatives]? Need was defined if a percentage greater than zero was entered under Converted Vehicles Not Designed as FD Apparatus.

Q24e. Number of engines/pumpers in service [that are] 30 or more years old. Need was defined if a number greater than zero was entered in this blank.

Sufficiency. There were not enough engines to equip enough stations (one engine per station), optimally located, to provide community coverage in accordance with NFPA standards and ISO formulas. This employed the formulas used in the Needs Assessment report, with different distance criteria for smaller vs. larger communities, as described in the report.

Table 5

Reported Needs vs. Awarded Grants – Vehicles and Facility Modification (Continued)

Q23d. Number of fire stations/Number equipped for exhaust emission control (e.g., diesel exhaust extraction). Need was defined if a number greater than zero was entered in this blank.

Source: USFA files on Fire Act grantees for "Firefighting Vehicles" and "Vehicles", and matching to USFA/NFPA Needs Assessment 2001 survey responses

20

Table 6

Reported Needs vs. Grant Amounts – Vehicles and Facility Modification

Community Size	Vehicle Percent of Dollars Granted	Need Any Question	Need Q6e	Need Q24e	Need Sufficiency	Facility Percent of Dollars Granted	Need Q23d
500,000 or more	1%	100%	0%	0%	100%	16%	86%
250,000 to 499,999	0%	NA	NA	NA	NA	16%	44%
100,000 to 249,999	4%	41%	0%	26%	41%	8%	53%
50,000 to 99,999	8%	65%	4%	10%	58%	8%	69%
25,000 to 49,999	10%	42%	2%	8%	40%	10%	69%
10,000 to 24,999	14%	70%	19%	28%	53%	9%	74%
5,000 to 9,999	22%	82%	31%	32%	63%	5%	79%
2,500 to 4,999	30%	85%	36%	39%	62%	3%	68%
Under 2,500	40%	88%	41%	46%	66%	1%	70%
Total	20%	80%	31%	36%	61%	7%	70%

NA – Not applicable because there were no such awards.

Note: Reported needs defined by indicated responses to questions. Need requires positive indication of need; blank answer to question is interpreted as no need. Percents in second and seventh columns are percent of all grant dollars to departments in that population stratum for which the grant was for the indicated resource, which here is vehicles and facility modification, respectively. Percents in third through sixth and eighth columns are (number of grant dollars to departments in that population stratum for the indicated resource where the grantee department reported a need on any of the specified questions)/(number of grant dollars to departments in that population stratum for the indicated resource). Vehicles and facility modification accounted for $191 million and $63 million, respectively, of the $960 million in grant money in the database.

Q. 6: What share (%) of your apparatus was [each of the listed alternatives]? Need was defined if a percentage greater than zero was entered under Converted Vehicles Not Designed as FD Apparatus.

Q24e. Number of engines/pumpers in service [that are] 30 or more years old. Need was defined if a number greater than zero was entered in this blank.

Sufficiency. There were not enough engines to equip enough stations (one engine per station), optimally located, to provide community coverage in accordance with NFPA standards and ISO formulas. This employed the formulas used in the Needs Assessment report, with different distance criteria for smaller vs. larger communities, as described in the report.

21

Table 6

Reported Needs vs. Grant Amounts – Vehicles and Facility Modification (Continued)

Q23d. Number of fire stations/Number equipped for exhaust emission control (e.g., diesel exhaust extraction). Need was defined if a number greater than zero was entered in this blank.

Source: USFA files on Fire Act grantees for "Firefighting Vehicles" and "Vehicles", and matching to USFA/NFPA Needs Assessment 2001 survey responses

22

Table 7
Reported Needs vs. Awarded Grants – Training

Community Size	Training Percent of Awards	Need Any Question	Need Q13a,b	Need Q14a,b	Need Q15a,b	Need Q16a,b	Need Q17a,b
500,000 or more	13%	46%	8%	8%	8%	31%	23%
250,000 to 499,999	14%	87%	7%	20%	27%	13%	87%
100,000 to 249,999	14%	82%	8%	16%	24%	43%	73%
50,000 to 99,999	12%	72%	9%	19%	23%	32%	58%
25,000 to 49,999	11%	82%	20%	15%	40%	38%	64%
10,000 to 24,999	11%	85%	29%	28%	51%	46%	61%
5,000 to 9,999	9%	88%	43%	34%	60%	61%	50%
2,500 to 4,999	8%	93%	53%	43%	64%	65%	46%
Under 2,500	7%	96%	68%	42%	63%	74%	47%
Total	9%	88%	40%	32%	52%	55%	55%

Note: Reported needs defined by indicated responses to questions. Need requires positive indication of need; blank answer to question is interpreted as no need. Percents in second column are percent of all grants to departments in that population stratum for which the grant was for the indicated resource, which here is training. Percents in third through eighth columns are (number of grants to departments in that population stratum for the indicated resource where the grantee department reported a need on any of the specified questions)/(number of grants to departments in that population stratum for the indicated resource). Training accounted for 1,335 of the 14,925 grants in the database.

Q13a. Structural firefighting. Is this a role your department performs?
Q13b. If yes, how many of your personnel who perform this duty have received formal training (not just on-the-job)? Need is defined if the answer to Q13a is Yes and the answer to Q13b is Most, Some, or None, and therefore not All.

Q14a. Emergency medical service (EMS). Is this a role your department performs?
Q14b. If yes, how many of your personnel who perform this duty have received formal training (not just on-the-job)? Need is defined if the answer to Q14a is Yes and the answer to Q14b is Most, Some, or None, and therefore not All.

Q15a. Hazardous materials (Hazmat). Is this a role your department performs?
Q15b. If yes, how many of your personnel who perform this duty have received formal training (not just on-the-job)? Need is defined if the answer to Q15a is Yes and the answer to Q15b is Most, Some, or None, and therefore not All.

Table 7

Reported Needs vs. Awarded Grants – Training (Continued)

Q16a. Wildland firefighting. Is this a role your department performs?

Q16b. If yes, how many of your personnel who perform this duty have received formal training (not just on-the-job)? Need is defined if the answer to Q16a is Yes and the answer to Q16b is Most, Some, or None, and therefore not All.

Q17a. Technical rescue. Is this a role your department performs?

Q17b. If yes, how many of your personnel who perform this duty have received formal training (not just on-the-job)? Need is defined if the answer to Q17a is Yes and the answer to Q17b is Most, Some, or None, and therefore not All.

Source: USFA files on Fire Act grantees for "Training" and "EMS Training", and matching to USFA/NFPA Needs Assessment 2001 survey responses

Table 8
Reported Needs vs. Grant Amounts – Training

Community Size	Training Percent of Dollars Granted	Need Any Question	Need Q13a,b	Need Q14a,b	Need Q15a,b	Need Q16a,b	Need Q17a,b
500,000 or more	6%	42%	11%	2%	11%	38%	15%
250,000 to 499,999	4%	80%	4%	11%	18%	7%	80%
100,000 to 249,999	5%	76%	5%	18%	23%	40%	72%
50,000 to 99,999	7%	71%	6%	15%	10%	25%	60%
25,000 to 49,999	6%	82%	13%	8%	40%	29%	71%
10,000 to 24,999	5%	82%	27%	29%	45%	45%	56%
5,000 to 9,999	3%	91%	41%	27%	66%	57%	55%
2,500 to 4,999	2%	92%	43%	37%	58%	66%	58%
Under 2,500	2%	97%	64%	46%	72%	76%	48%
Total	4%	80%	24%	21%	40%	43%	58%

Note: Reported needs defined by indicated responses to questions. Need requires positive indication of need; blank answer to question is interpreted as no need. Percents in second column are percent of all grant dollars to departments in that population stratum for which the grant was for the indicated resource, which here is training. Percents in third through eighth columns are (number of grant dollars to departments in that population stratum for the indicated resource where the grantee department reported a need on any of the specified questions)/(number of grant dollars to departments in that population stratum for the indicated resource). Training accounted for $37 million of the $960 million in grant money in the database.

Q13a. Structural firefighting. Is this a role your department performs?
Q13b. If yes, how many of your personnel who perform this duty have received formal training (not just on-the-job)? Need is defined if the answer to Q13a is Yes and the answer to Q13b is Most, Some, or None, and therefore not All.

Q14a. Emergency medical service (EMS). Is this a role your department performs?
Q14b. If yes, how many of your personnel who perform this duty have received formal training (not just on-the-job)? Need is defined if the answer to Q14a is Yes and the answer to Q14b is Most, Some, or None, and therefore not All.

Q15a. Hazardous materials (Hazmat). Is this a role your department performs?
Q15b. If yes, how many of your personnel who perform this duty have received formal training (not just on-the-job)? Need is defined if the answer to Q15a is Yes and the answer to Q15b is Most, Some, or None, and therefore not All.

Table 8

Reported Needs vs. Grant Amounts – Training (Continued)

Q16a. Wildland firefighting. Is this a role your department performs?
Q16b. If yes, how many of your personnel who perform this duty have received formal training (not just on-the-job)? Need is defined if the answer to Q16a is Yes and the answer to Q16b is Most, Some, or None, and therefore not All.

Q17a. Technical rescue. Is this a role your department performs?
Q17b. If yes, how many of your personnel who perform this duty have received formal training (not just on-the-job)? Need is defined if the answer to Q17a is Yes and the answer to Q17b is Most, Some, or None, and therefore not All.

Source: USFA files on Fire Act grantees for "Training" and "EMS Training", and matching to USFA/NFPA Needs Assessment 2001 survey responses

Table 9

Reported Needs vs. Awarded Grants –

Wellness/Fitness and Fire Prevention Programs

Community Size	Wellness/Fitness Percent of Awards	Need Q18	Fire Prevention Percent of Awards
500,000 or more	11%	27%	7%
250,000 to 499,999	13%	57%	8%
100,000 to 249,999	13%	64%	6%
50,000 to 99,999	8%	70%	8%
25,000 to 49,999	9%	55%	6%
10,000 to 24,999	6%	65%	4%
5,000 to 9,999	3%	66%	2%
2,500 to 4,999	2%	77%	1%
Under 2,500	2%	67%	1%
Total	4%	64%	3%

Note: Reported needs defined by indicated responses to questions. Need requires positive indication of need; blank answer to question is interpreted as no need. Percents in second and fourth columns are percent of all grants to departments in that population stratum for which the grant was for the indicated resource, which here is wellness/fitness and fire prevention, respectively. Percents in third column are (number of grants to departments in that population stratum for the indicated resource where the grantee department reported a need on any of the specified questions)/(number of grants to departments in that population stratum for the indicated resource). Wellness/fitness and fire prevention accounted for 627 and 397, respectively, of the 14,925 grants in the database.

Q18. Does your department have a program to maintain basic firefighter fitness and health (e.g., as required in NFPA 1500)? Need was defined by a No answer.

Source: USFA files on Fire Act grant recipients and matching to USFA/NFPA Needs Assessment 2001 survey responses

Table 10
Reported Needs vs. Grant Amounts –
Wellness/Fitness and Fire Prevention Programs

Community Size	Wellness/Fitness Percent of Dollars Granted	Need Q18	Fire Prevention Percent of Dollars Granted
500,000 or more	12%	13%	4%
250,000 to 499,999	20%	74%	3%
100,000 to 249,999	11%	73%	2%
50,000 to 99,999	8%	80%	5%
25,000 to 49,999	6%	50%	4%
10,000 to 24,999	4%	68%	3%
5,000 to 9,999	1%	58%	2%
2,500 to 4,999	1%	76%	1%
Under 2,500	0%	66%	0%
Total	4%	62%	2%

Note: Reported needs defined by indicated responses to questions. Need requires positive indication of need; blank answer to question is interpreted as no need. Percents in second and fourth columns are percent of all grant dollars to departments in that population stratum for which the grant was for the indicated resource, which here is wellness/fitness and fire prevention, respectively. Percents in third column are (number of grant dollars to departments in that population stratum for the indicated resource where the grantee department reported a need on any of the specified questions)/(number of grant dollars to departments in that population stratum for the indicated resource). Wellness/fitness and fire prevention accounted for $38 million and $22 million, respectively, of the $960 million of grant money in the database.

Q18. Does your department have a program to maintain basic firefighter fitness and health (e.g., as required in NFPA 1500)? Need was defined by a No answer.

Source: USFA files on Fire Act grant recipients and matching to USFA/NFPA Needs Assessment 2001 survey responses